MORSE CODE DECIPHERED

A Meaning Behind The Code

R. Lindo

Wellside Press

Table Of Contents

CHAPTER 1: INTRODUCTION

AUTHOR'S NOTE

This course is intended to hold a student's interest in Morse Code by instilling or restoring the precious advantage of self-confidence. The rapid progress possible with this course does just that. It slices away many of hours of repetitious practice along with the notion that one has to practice a single letter thousands of times before it can be absorbed. With derived meanings, vivid associations and the interesting steps offered, Morse Code becomes enjoyable and conquerable.

This booklet is perfect for either for self-study or a classroom type presentation. It also works great if used to supplement any other course in Morse Code, or just to increase speed, or to polish up on proficiency. The most enthusiasm usually comes from people who have tried to learn the code before but for whatever reasons gave up on it. A quick revision is all it will take to regain proficiency.

With old and conventional methods, the code is taught to be digested raw; First the student must memorize by sight or audio, and later the code must somehow be instantaneously recalled. This method can be one giant step into confusion and frustration, because there are twenty-six of these weird and strikingly similar code sounds for letters alone, not including numbers and punctuation marks. Sheer memorization can become a little overwhelming.

THIS MANUAL

This method though very simple is very effective. It's just a very clever way of learning. The manual uses a combination of two methods. I call them the "phonetic link" and the "code sound phrase" method.

It is both a study guide and a review guide in one. Everything presented was done with that in mind. It was written with the idea of day one, page one. It starts simple and it stays simple.

WHY THE CODE IS IMMORTAL

Morse Code is elegantly simple yet highly effective. A flashlight is its simplest form of transmission and, providing there is enough visibility, is as effective as if by any sophisticated radio transmission. Morse can even be transmitted by knocking or scratching on a wall, or through physical contact.

For radio wave transmission Morse requires only relatively cheap unsophisticated equipment and very little power to reach very long distances because it does not need a constant carrier wave as in the case of FM or with voice transmission (telephony). It also takes up less frequency space. This is important because the frequency bands grow more crowded each day.

Unlike the spoken word with its close intricacies which have to be heard distinctively to be understood, Morse has two easily distinguishable syllables, dit(.) and daah(-). That's all. These two syllables will usually be impervious to atmospheric interference.

NASA has even put satellites in orbit exclusively for carrying amateur radio. The system is such that a person can have his or her very own little inexpensive communication station and talk with random people all over the globe thereby spreading a little international goodwill. The F.C.C. says in effect, "Talk all you want but your station should not be used as a means to derive profit". This is known as the Amateur Service.

Morse Code is also a vital aid in aircraft navigation as it is used to positively identify ground stations or navigation aids (particularly VORs and ILSs) which are either homed in on or used as a reference for a position fix. The name of the ground station is abbreviated and transmitted in Morse Code over the ground station carrier frequency.

The medium or method of transmission or signaling depends on your own ingenuity. It is quite easy to communicate using Morse Code through physical contact. All that's required is a way to express a long daah(-) and a way to express a short dit(.). This ingeniously simple way to encode letters and thus words, will undoubtedly continue to find useful existence.

CHAPTER 2: FAMILIARIZATION

ANNOUNCING THE TWO AND ONLY TWO SYLLABLES

The dit or the short sound, also called the dot. The daah or the long sound, also called the dash. The way each is written is symbolic to the duration of the sound or signal that it represents i.e., a tiny dot for the short sound and a line for the long sound.

Each letter of the alphabet is a particular arrangement of short (dit) and long (daah) sounds. A letter can vary from one to four elements. Numbers, however, are always five elements and follow a set pattern. Punctuation marks are usually six elements. Letters, numbers, and punctuation marks are collectively called characters. We'll start with letters first because everything that applies to letters also applies to the other characters.

THE TWO MAIN METHODS OF TRANSMISSION

(AURAL OR VISUAL)

The code is generally transmitted by two methods: aural or visual. When the characters are converted to or represented by sound it is known as the aural method. With this method, the length of the sound is varied to differentiate a dit from a daah. A short sound which sounds like a quick knock, represents the small dit(.). A long sound represents the daah(-).

If the code is transmitted by using a light source, such as a flashlight, it's called the visual or sight method. To represent the short dit, the light source is turned on for just an instant. Instead of a long sound for the daah, the light source is simply kept on for a longer duration of time.

For some people, it is easier to learn by ear rather than by sight because a rhythm is usually easier to recognize if heard rather than if seen. Choose the method that works best for you including a combination of both. Even experts are often heard to hum the sound equivalent, most times unconsciously, when transmitting or deciphering code sent by the visual method.

This is because the syllables, dit and daah, are named from their actual sound or the aural method. The aural method can have the advantage of the student not having to keep his or her eyes fixed to a light source and is thus able to look at what he or she is copying.

RATIO AND SPACING (PROPORTIONATE PAUSES)

In order to maintain a standard flow, the long sound should be three times the duration of the short. This three-to-one duration ratio (also termed as a "weight" of three) is held true for all speeds. The reason for weight is not for pinpoint consistency but to keep somewhat of a standard recognizable rhythm. For a beginner, a quick knock or touch for the dit and approximately a three second duration for the daah is good enough.

As you get more advanced, a separation period of one dit between letters and a separation period of one daah between words is ideal. This separation rarely ever needs to be practiced in the early stages because it usually naturally falls into place.

Allow yourself enough and proportionate spacing between your letters, words, and sentences in that order. The pause between words should be longer than the pause between letters, and the pause between sentences even longer. These are the same pauses that are observed in regular speech. They should get shorter as you get more proficient. Don't try to go too fast, your speed will increase naturally.

IN KEEPING WITH THE TRUE SOUND

The piercing "t" in a dit is particularly distinct at the end of a code sound, however, in any other position, the "t" is absorbed by the next sound. It is almost as if the "t" is chopped off, making the dit sound like a sharp "di". For example, an R actually sounds like di' <u>daah</u> dit (.-.).

With this in mind and from here onward, when applicable, a dit will be written as a di to match the actual rhythmic sound as shown above and the <u>daahs</u> will be underlined for easier recognition. Secondly, even though that's what they are and look like, it is not correct to refer to the syllables as the dot or the dash because that is not what they sound like in actuality. For example, try not to call an N(-.) dash-dot. It's really <u>daah</u>-dit. Try to call the letters by their true sound.

HOW TO HEAR AND RECOGNIZE

The key to hearing the code and recognizing the characters is knowing how to listen. To an untrained person, everything sounds the same. Even if told about the long sound and the short sound, an untrained person will not be able to tell them apart.

In the familiarization stage, focus on the sharp piercing abrupt dit and listen for detail. The faster the code is transmitted, the sharper the dits become. If you can recognize the dits then you'll surely hear the daahs as they are three times as long. If you can recognize the dits and daahs then the spacing between the letters and words will seem very obvious. Keep in mind though that eventually you will learn to recognize the overall rhythm. The speed at which each character is transmitted will vary but the rhythm itself will stay the same.

Know that the code seems to change at about 10 words per minute when a character starts to become a "continuous flow". At this speed there no longer seems to be a space between the different elements in a character. This lack of space causes the dit's distinctive t's to disappear. (As described in the previous section, "IN KEEPING WITH THE TRUE SOUND"). The dits suddenly lose the piercing t's and change to di's. This can throw off the rhythm and can make the transition beyond this speed a little difficult for some unless they are made aware.

Compare dit-daah (8 wpm) with dididaah (12 wpm). The two di's in dididaah at 12 wpm are usually, and easily, confused as being a single dit because with the faster speed, the characteristic t's disappear. In addition, there are now two dits snugly fit into the same time period previously used for just one. This slight distortion of rhythm makes decoding at 8 wpm subtly but significantly different from decoding at 12 wpm, especially if you've gotten used to decoding at 8 wpm.

The Farnsworth method of transmission speeds through the characters but increases the space between characters. There is a high character speed but a lower overall send speed. This gives the characters the "continuous flow" effect but without a high wpm rate. Using the Farnsworth method from day one has been one way to get around the difficultly with the transition.

Some people believe that it is best to start out with the Farnsworth method, but the characters may be too fast for some beginners. In addition, merely explaining the slight but significant rhythmic difference or shift that occurs at about 10 wpm because of the dits, usually clears up the problem that some people have of being stuck at about this speed.

I recommend taking some time to get used to the codes first and then deciding what works best for you. About 10 wpm or when the dits and daahs start to flow into each other, listen for di's and remember that the rhythms are going to be made up of di's. Don't listen for dits.

I use the didi as a reference base. The same also goes for the daah daah. I recognize a didi didi when I hear two groups. In other words, when I hear the didi twice. A divididit is thus a single group with one extra, and a didi didi dit is two groups and one extra. This way the extra dit (or daah) always stands out.

Without trying to decode, as this is not meant to be a test of your deciphering ability but rather an effort to develop and improve sequence recognition, listen for consecutive elements i.e., a double dit or a double daah. Listen as if to point out these double elements (didi or daah daah) to someone else. Doing this will give you all you need which is just a sense of the rhythm.

At first one might be tempted when decoding to convert the code heard into dits and daahs as they are written on paper. This is acceptable in the early confidence building familiarization stage, but ultimate objective is the recognition of the rhythm and not so much the recognition of individual dits and daahs. Entirety is the key. Entirety is everything. Entirety is your goal.

Code rhythms can be associated with words or phrases that they sound like; For example the code for Y(ankee) which is -.-- (daah-di-daah-daah) seems to say the phrase, "Yankee doodle", and the code sound for X(-ray) which is -..- (daah-di-di-daah) seems to say "only the bones". I call these "code sound phrases".

When you have become comfortable with recognizing dits from daahs, the code sound phrases and the code in general, go straight to deciphering at 16 wpm. In other words, with this method you can start out decoding at about 16 wpm. With regards to your send speed, you can build your send speed at your rate.

Sometimes when deciphering you'll miss a letter or two entirely and it will then be helpful if you remember the number of elements in the letter. See "THE NUMBER OF ELEMENTS" in the "PRACTICE GROUPS" section. As you get better though, you will also be able to fill in a missing letter or two from just looking at the other letters in the word.

CHAPTER 3: THINGS TO KNOW BEFORE YOU START

READ THROUGH THE ENTIRE BOOKLET

This ensures that no important points are missed before starting.

BEGIN SLOWLY

Make your daahs (the long sound) extra-long. For making a daah if using a light hold it on, with a switch hold it down for about three seconds. The dits (shorts) can just be a quick knock or touch.

NAME THE TUNE, PLAY THE TUNE

Transmitting code is, in a sense, just playing a particular tune for a particular letter. Deciphering is thus naming the letter behind the tune. As overwhelming as it may appear initially, that is all it is. With the 10-wpm shift in mind, it is worth remembering that regardless of how fast the code is transmitted beyond that, a letter keeps the same rhythm.

It is probably very easy to recognize one of your favorite songs after hearing just the first few seconds of it. Some people are even able to recognize an instrumental if allowed to hear a very short piece of it. Morse Code is no different. As humans, we all have a sense of rhythm recognition. Once you get the feel, don't be alarmed by your own ability.

TRUST YOUR GUT

It is a sure sign of learning when you hear a letter, and you are able to tell what it is just by the rhythm of it alone. It means that you are on the right track. As you become more proficient, each letter will seem to simply say its own name. In time, each will eventually become a particular and distinct piece of music.

USE A PROBLEM LIST

To optimize your time, and very effectively so, make a 'problem list' of the letters that you have some amount of difficulty with. There are always going to be a few of them that deserve, and should get, most of your attention. Don't spend as much time on the easy ones or those you already know. Remember that the most that you can do for any letter or character is learn it. After you learn it, move to another.

The problem codes should be the heart of any revision with the objective being to conquer the list. Write down anything that helps you to remember difficult codes and use it. Take the codes off the list as you learn them or make up a new list.

When you want to increase your speed, the list should reflect the characters that you are a little slower on or those characters that stall you a little sometimes.

REVIEW REGULARLY

This will make sure you do not have enough time to forget anything and will keep codes from returning to the problem list. There is an entire section on how to review.

CHAPTER 4: THE STANDARD PHONETIC (OR WORD) ALPHABET

ALPHA, BRAVO, CHARLIE

We are about to go from the usual letter alphabet of saying the alphabet as A, B, C, D etc. to saying the phonetic or word alphabet, wherein letters are replaced with entire words. The standard phonetic alphabet starts out Alpha, Bravo, Charlie, Delta, Echo. These same exact standard words are used across the board in radio communications globally.

Letters called by name sound too much alike and are almost impossible to distinguish sometimes; For example, it's easy to confuse E with B, C, D, G, P, T, V or Z. Saying the standard word Echo, for the letter E, eliminates the possibility of misunderstanding.

This will be especially useful with your international contacts because some people won't be familiar with the names of letters in English as they are only used to hearing the standard phonetics.

THE LINK

A meaning is created for, blended into, and stands behind each standard phonetic word, (and thus the particular letter). This meaning provides a visual image which acts as a memory aid or link. It serves as an intermediate step and makes the codes digestible.

Furthermore, code sounds, for example, di-di-daah (..-) which is the code for U(niform), are linked to close sounding 'code sound words' or 'code sound phrases' which are also associated to the standard phonetic word.

In this case with Uniform, the code sound word is also the standard phonetic word because the code sound di-di-daah (..-) seems to distinctively say, and sounds exactly like the word "U'ni'form".

In other words, letters are read and treated as their complete standard phonetic word, and this word evokes the code. Learn the standard phonetic word for each letter and with this method, that alone will be a significant step towards conquering Morse Code.

Although not immediately obvious as say "Charlie Charlie", the letter represented, except for X(-ray), always appears in its code sound phrase; For Example, Bravo (only the best), Hotel (Ho'li'day Inn), Juliet (The Julie Jay), Lima (The Lima El), Romeo (The roses), and Whiskey (The water).

CHAPTER 5: MORSE CODE DECIPHERED (EXPANDED EXPLANATIONS)

THE LETTERS

.- ALPHA (di-daah)
Hieroglyphics never got any better than this. The word and the code both fittingly represent the very beginning as in the first letter of the alphabet. The dit represents a sort of start or spark and the daah represents a form of continuation or progression. The code sound is "Al'phahh".

-... BRAVO (daah-di-di-dit)
As in the applause received after a performance, it's an expression of commendation or admiration for a worthy accomplishment. The code sound says the phrase "Only the be'st".

-.-. CHARLIE (daah-di-daah-dit)
The code sound says "Charlie, Charlie". See Mike.

-.. DELTA (daah-di-dit)
The code sound describes swimming in a delta, "Delta dip". The dip also sounds like, and can be used as a reminder of, the dits at the end of the code sound. Can also be remembered as the "Delta dits".

. ECHO (dit)
This is the shortest code sound, and the easiest to transmit. The single dit can easily be associated with a sonar radar return. The code sound word "Ec" can also be considered to be the original sound that's subsequently echoed, or just as short for the standard phonetic word.

..-. FOXTROT (di-di-daah-dit)
The word comes from that old American two-step ballroom dance. Everyone was always well dressed at a ball. It was certainly not like going to the discotheque which appeared later. The code sound phrase is the rhythm to this crazy boogie and says "Do the Foxtrot". Also note that the Foxtrot was a two-step dance so it starts with two dits, and is mostly all dits, in keeping with the rhythm of the music.

--. GULF (daah-daah-dit)
The code sound says "Gulf Pie'rit", in other words Gulf Pirate".

.... HOTEL (di-di-di-dit)
The code sound says "Ho'li'day Inn".

.. INDIA (di-dit)
The code sound is "Indi".

.--- JULIET (di-daah-daah-daah)
The code says "The Julie Jay".

-.- KILO (daah-di-daah)
The code word says "Kilogram". Kilo is a precise prefix. It doesn't mean 1001 or 999. It means 1000. This precision is represented by the dit between the two daahs which resembles the reading on a scale. It's another hieroglyph like the letters (.-) Alpha and November (-.)

.-.. LIMA (di-daah-di-dit)
The code sound says, "The Lima el". Lima is the capital of modern-day Peru where the first South American civilization appeared, the vast Inca empire.

-- MIKE ***(MA'IKE)*** (daah-daah)
The code sound simply says "Mike, Mike", (See Charlie) or when you get way up there at about 20 wpm, the code sound then seems to say "Ma'ike" for Mike.

-. NOVEMBER (daah-dit)
The code sound says Nooo'vem. This is the hieroglyphic and conceptual opposite of Alpha. November signifies an ending or discontinuity as in the concept that summer is over by this time in the northern hemisphere (and that winter is over by this time in the southern).

--- OSCAR (daah-daah-daah)
The code sound says, "Oscar O".

.--. PAPA (di-daah-daah-dit)
The code sound says "The paah paah Pee".

--.- QUEBEC (daah-daah-di-daah)
This is the largest of the Canadian provinces. It's got much more than its fair share of large rivers, bays and lakes and it's even got its own gulf, the Gulf of St. Lawrence. The code sound seems to say "Quebec, le Q" or "Que'bec is Q".

.-. ROMEO (di-daah-dit)
The code sound says, "The roses" or "The Ro'mi".

... SIERRA (di-di-dit)
The code sound is "Si'er'ra". It's Spanish for mountain as in Sierra Nevada, "snowy saw-toothed mountain range", Sierra Leone "lion mountain", and Sierra Madre "mother mountain".

- TANGO (daah)
This code sound simply says "tang". The Tango, the once popular Latin American beat was so named because of its distinct flavor, or tang.

..- UNIFORM (di-di-daah)
The code sound says exactly that "Uniform".

...- VICTOR (di-di-di-daah)
Beethoven started his 5th symphony with this code sound, "da da da daah". It's the perfect thing to listen to in celebration of victory and he is the perfect example of a true victor. He overcame trying times all throughout his life. He was not a child prodigy like Mozart and as an adult had to prove his worth. His genius spurred from effort, persistence, and determination. This code sound is a salute to Beethoven and is victor music.

.-- WHISKEY (di-daah-daah)
The word whiskey is from the Gaelic usage "beatha", the "water of life". The native Americans called it firewater. The code sound seems to say "the wa'ter".

-..- X-RAY (daah-di-di-daah)
This very technical code sound describes what you see with X-rays, "only the bones".

-.-- YANKEE (daah-di-daah-daah)
The code sound is made by a rhythmic yanking type motion or attitude, yanking on the key, being mostly all daahs. It plainly says, "Yankee doodle".

--.. ZULU (daah-daah-di-dit)
The most famous king of the African Zulus, the code sound says his name, "Sha'ka Zulu".

PUNCTUATION MARKS

.-.-.- The period (.) consists of (A)lpha written three times in succession with (A)lpha being the first letter of the alphabet and it signals that a new sentence is about to start.

..--.. The question sign (?) which is also used to request a repetition of a transmission that's not clearly understood, can be remembered as a request to expand on content thus the long daahs in the middle. The long daahs in the middle are the content that you want expanded or explained.

--..-- The comma (,) like the dits between the long daahs is used to separate lengthy thoughts.

-..-. The slash (/) is one half of an X or in other words, an abbreviated X. For the abbreviated X, the code is an X (-..-) followed by a period, [X.].

.-..-. Quotation marks (**"**) are put around and enclose, *"xactly"* what's said thus the X(ray) is used to represent the "xact" content enclosed by dits, one on either side of the [.X.].

-.-.-. The semicolon (;) consists of (N)ovember written three times and signifies the end (like the end of a year) of a main thought.

THE NUMBERS

Numbers always have five elements. This helps to differentiate them from letters which are limited to a maximum of four. They are somewhat like Roman numerals, but they are still supposed to be learned by rhythm although not necessarily by just rhythm alone.

The lower numbers (1 to 5) start with a dit. The higher numbers (6 to 0) start with a daah. One method is to simply count the number of whichever syllable (dit or daah) that the number starts with, with the smaller dit being equal to one. A daah, as it is literally and symbolically bigger is equal to the number of dahs counted plus five if the number starts with a dah; In other words, if a number starts with a dah then there is an invisible 5.

The Numbers

1 2 3 4 5	6 7 8 9 0
The Lower Numbers	**The Upper Numbers**
1 .----	6 -....
2 ..---	7 --...
3 ...--	8 ---..
4 ....-	9 ----.
5	0 -----
1 dit = 1	1 daah (+5) = 6
2 dits = 2 etc	2 daahs (+5) = 7 etc

THE TRICK TO NUMBERS (SEEING A NUMBER BOTH WAYS)

Dit = one or minus ten
Daah = plus five or minus five

Dits are always equal to one wherever they appear in a number. The one will either be added or subtracted depending on where the dits appear in the number. If the number starts with a dit then just add them up. All you do is count the dits as in (..---) which in this case is the number two.

If the dit is towards the end of a number as in (---..) then just subtract the number of dits from ten. In this case the number is eight as two from ten equals eight.

The daah, when it appears first is equal to the number of daahs counted plus five, as in (--...) which is the number seven because there are two daahs plus five. If the daah appears towards the end as in (....-) then its value becomes the amount of daahs counted subtracted from five. That's because if a number starts with a dit then it can't be any bigger than five. In this case the number is four. There is one daah and one subtracted from five is four.

If you look closely at the last number again, you'll notice it could have also been decoded by simply counting the dits, as it starts with a dit. The trick to numbers is what was just demonstrated and it is just being able to see a number both ways, from the back or the front, and then adding or subtracting depending on which is easier.

A dit is (1). Add up the dits if it appears first, subtract from ten if it does not. A daah is (5). Add five if appears first, subtract from five if it does not.

THE CIRCULARS (Q, O, 0 AND 6, 8, 9)

I call Quebec (--.-), Oscar (---) and the number zero (0) (-----) circulars as they are all circles and have very similar code sounds. Notice the characteristic consecutive daahs. They all start with at least two daahs and are mostly all daahs. The other circulars are all the numbers with circular portions (6, 8 and 9) and they too also start with daahs.

Quebec	--.-	6 (six)	-....
Oscar	---	8 (eight)	---..
0 (zero)	-----	9 (nine)	----.

CHAPTER 6: HOW TO LEARN THE CODE SOUNDS AND PRACTICE ANYWHERE

THE PHONETIC WORDS: 'SILENT PHONETICS'

All it takes to learn the standard phonetic words, thus learning the standard phonetic alphabet, is becoming familiar with them. You can practice by covering the words below with just the first letters in view and then trying to remember the rest of the word. Keep at it until you've got them all. You can start by learning how to say your name in the Standard Phonetic Alphabet.

Alpha .-	Juliet .---	Sierra ...
Bravo -...	Kilo -.-	Tango -
Charlie -.-.	Lima .-..	Uniform ..-
Delta -..	Mike --	Victor ...-
Echo .	November -.	Whiskey .--
Foxtrot ..-.	Oscar ---	X-ray -..-
Gulf --.	Papa .--.	Yankee -.--
Hotel	Quebec --.-	Zulu --..
India ..	Romeo .-.	

When you've got the feel of it, find a book, a newspaper or anything interesting and current. To practice, read out the letters in your mind in phonetics as if for someone else to copy down; For example, the word Morse becomes Mike, Oscar, Romeo, Sierra, Echo. We'll proceed from here onward calling the letters by their phonetic names.

THE CODE SOUNDS: 'SILENT CODE PRACTICE'

Because the code sounds are directly blended in with the particular phonetic word, it is best to become somewhat familiar with both at about the same time. If there is anything readable in sight including this manual, anywhere you are, translate the letters to Morse Code as if sending for someone else to receive (similarly to what was done to learn the phonetic words). A once normal exit sign now becomes:

Echo	dit	*"Ec"*
X-ray	daah-di-di-daah	*"Only the bones"*
India	di-dit	*"In'di"*
Tango	daah	*"Tang"*

THE TRICK TO TRANSMITTING AT RAPID SPEEDS

PHYSICAL REACTION

The code sound should be considered as just a different way of writing a character; Be it a letter, punctuation mark, or number. Each letter phonetic is meant to initiate an instantaneous physical response or reflex reaction, and it's this reaction that's transmitted.

For example for Echo (.), you should think of yourself as having to transmit an echo; Saying "only the bones" for X-ray (-..-); Saying "Oscar O" for the letter O (---) as if Oscar owns it; or simply saying "Uniform" for Uniform.

Saying Charlie (-.-.) twice; saying Si'er'ra (...); Specifically naming the Ho'li'day Inn for Hotel (....); playing Victor in salute to Beethoven. Just to name a few.

Think reflex and look over the phonetics for the response that each is meant to trigger and practice trying to decipher and to transmit by using just the code sound words and code sound phrases. Remember to call the letters by their phonetic names.

You will soon acquire a certain proficiency when sending wherein you'll hear the code sound for the letter that you are going to send right before you have to send it without having to think. It means that you are starting to break away from the phonetics and going directly from letter to code sound. This is the goal. The codes will eventually become almost the same as regular speech, instantaneous and requiring no effort or forethought.

CHAPTER 7: NOW THAT YOU KNOW MORSE CODE

THE INTERNATIONAL DISTRESS SIGNAL

It is called the SOS signal, generally considered to be the abbreviation for “Save Our Souls”. It is transmitted if threatened by grave and imminent danger. The alternating three short dits and three long daahs is very distinct. Sierra (...), Oscar (---), Sierra (...) or di-di-dit, daah-daah-daah, di-di-dit. Learning the distress signal should also help with learning and remembering the letters Sierra and Oscar. There is no pause between the letters but there is a pause between each sequence.

SOS

Sierra	...
Oscar	---
Sierra	...

LEARN YOUR NAME IN MORSE CODE

The next thing that you should probably learn to send after the distress signal is your name. You should be able to send your name instantly if called upon. Learn your name in code and learn it well. Later you can start on the names of your favorite people.

CHAPTER 8: PRACTICE GROUPS

The best groups are those that you arrange yourself based on your experiments or preference. Practice groups are always nice exercises for fun and proficiency but if you have learned most of the codes then don't waste too much of your time with letters, or punctuation, or numbers, arranged in all sorts of groups. Work the problem list instead. You'll know yourself which letters you need to learn a little better.

There are millions of ways that you can group the codes but there are two basic groups that are very popular. The first is the single syllable and the other is based on the number of elements in the code.

GROUP 1: SINGLE SYLLABLE (ALL DITS OR ALL DAHS)

The simplest, and most basic of all groups, is the single syllable. It is a grouping of the codes that are all dits or all daahs. A little catch phrase or word to remember "eish" group is shown in quotation marks. *Eish* and *TMO* can be remembered as *"Each Team"* of all dits or all dahs.

EISH *"each"*

. Echo	dit	*"Ec"*
.. India	di-dit	*"Indi"*
... Sierra	di-di-dit	*"Si'er'ra"*
.... Hotel	di-di-di-dit	*"Ho'li'day Inn"*

TMO *"Team O"*

- Tango	daah	*"Tang"*
-- Mike	daah-daah	*"Mike Mike"*
--- Oscar	daah-daah-daah	*"Oscar O"*

ABOUT DERIVATIVES

A derivative is an outgrowth of a particular code. You have already been introduced to a simple set of derivatives in the previous "SINGLE SYLLABLE" group of all dits or all dahs, however, there is no reason to study a long series of what letters turn into what other letters because each letter is supposed to be learned as a complete whole. For example it is not necessary to study that a T(-) turns into an N(-.) if you add a dit or into an M(--) if you add a daah.

The groups presented are some of those groups that beginners seem to typically automatically construct. Trying to learn code by splitting code letters into other smaller letters is not recommended. For example, the letter Q should not be memorized as TK or MA as this creates an undesirable disruptive second step. The presented derivative lists are mostly helpful and useful for missed letters, rhythm development and recognition. Splitting is only generally accepted for Q signs and Prosigns because those are simple enough.

IF YOU MISS A FEW LETTERS

If you miss a few letters, write in what you think it sounded like even if is just a wild guess. Usually a wild guess, even if not correct, will fall within the sequence of derivatives, thus making some derivatives good to know. The single syllable groups of "eish" and "team o" which consist of codes that are all dits or all daahs are the best examples of good derivatives.

If you miss a letter when practicing with a friend, ask him or her to repeat the letter until you figure out what it is. Try not to pass them up. Knowing the two basic groups in the early stages does wonders for self-confidence. The way to learn is by gradually overcoming and extending your limitations. It's okay to miss a few letters. As a matter of fact, that's exactly how to increase your decoding speed; increase the speed of the transmission to a little above the speed that you feel comfortable with.

Try not to fall into the trap of trying to figure out a single missed letter as you miss entire words. A single letter will be obvious from looking at the entire word in the same way that a single word will be very obvious from looking at the entire sentence. There is usually enough time after a message to fill in blank spaces.

Bear in mind that some exchanges will contain errors. Don't freeze up if you transmit a letter that's not really a letter, and don't be too bothered when decoding if you miss something that seemingly had all the appearances of a valid letter. Keep in mind that no one is perfect so have a little patience with yourself and keep trying and keep at it.

GROUP 2: BY NUMBER OF ELEMENTS

You'll notice that the letters used most frequently have shorter codes. Uncommon letters were assigned the longer codes. For example, the letter E(cho) which is one of the vowels, is (.), but Z(ulu) is (--..). You may also notice that there are no vowels in the codes that have four elements. This was done after Samuel F.B. Morse chose the original codes so that it would be a little easier to speed through sentences. The following grouping is good to know because at times you will miss a code but still be able to tell how many elements were in it. They are ET, MAIN, GUK WORDS (guk words) and everything else.

(1) ONE ELEMENT (ET)

.	Echo	dit	*"Ec"*
-	Tango	<u>daah</u>	*"Tang"*

(2) TWO ELEMENTS (MAIN)

--	Mike	<u>daah</u>-<u>daah</u>	*"Mike Mike"*
.-	Alpha	di-<u>daah</u>	*"Al'phahh"*
..	India	di-dit	*"In'di"*
-.	November	<u>daah</u>-dit	*"Nooo'vem"*

(3) THREE ELEMENTS (GUK WORDS)

--.	Gulf	daah-daah-dit	*"Gulf pirate"*
..-	Uniform	di-di-daah	*"Uniform"*
-.-	Kilo	daah-di-daah	*"Kilogram"*
.--	Whiskey	di-daah-daah	*"The water"*
---	Oscar	daah-daah-daah	*"Oscar O"*
.-.	Romeo	di-daah-dit	*"The roses"*
-..	Delta	daah-di-dit	*"Delta dip"*
...	Sierra	di-di-dit	*"Si'er'ra"*

(4) FOUR ELEMENTS (BCFHJLPQVXYZ)

This group is made of the remaining letters being B,C, F, H, J, L, P, Q, V, X, Y and Z. The four groups in review are ET, MAIN, GUK WORDS (guk words) and then everything else.

CHARACTER FIRST OR CODE FIRST IS YOUR CHOICE

Sometimes in a practice group the code appears first and is followed by the character. At other times, the character may appear first. The order is not important as it is always a good idea to sometimes do the groups in reverse. For example, for Romeo (.-.), a person can be asked to send the code for R or to decipher the di-daah-dit. This balances sending and receiving skills when you are learning the code. Problem characters will be difficult to transmit as well as to decode.

CHAPTER 9: A CONTRACTION OF THE CODES

CONTRACTED LETTERS

.-	**ALPHA** "Al'phah" (di-daah) The hieroglyphic genesis
-...	**BRAVO** "Only the best" (daah-di-di-dit). For figuratively planting the flag on the mountain top
-.-.	**CHARLIE** "Charlie Charlie" (daah-di-daah-dit)
-..	**DELTA** "Delta dip" (daah-di-dit)
.	**ECHO** "Ec" (dit) Short for the phonetic word.
..-.	**FOXTROT** "Do the Foxtrot" (di-di-daah-dit). The rhythm of old two step American ballroom dance.
--.	**GULF** "Gulf pirate" (daah-daah-dit)
....	**HOTEL** "Ho'li'day Inn" (di-di-di-dit)
..	**INDIA** "Indi" (di-dit)
.---	**JULIET** "The Julie Jay" (di-daah-daah-daah)
-.-	**KILO** "Kilogram", "Kee'luh'gramm" (daah-di-daah)
.-..	**LIMA** "The Lima El" (di-daah-di-dit)
--	**MIKE** "Mike Mike" (daah-daah) or "Ma'ike"
-.	**NOVEMBER** (daah-dit) "Nooo'vem". The hieroglyphic and conceptual opposite of Alpha.
---	**OSCAR** "Oscar O" (daah-daah-daah

.--.	**PAPA** "The pahpah Pee" (di-daah-daah-dit)
--.-	**QUEBEC** "Quebec le Q" (daah-daah-di-daah)
.-.	**ROMEO** "The roses" (di-daah-dit)
...	**SIERRA** "Si'er'ra" (di-di-dit)
-	**TANGO** "Tang" (daah)
..-	**UNIFORM** "Uniform" (U'ni'form) (di-di-daah)
...-	**VICTOR** "da da da daaww" (di-di-di-daah). Beethoven's fifth, salute to Beethoven.
.--	**WHISKEY** "The water" (di-daah-daah)
-..-	**X-RAY** "Only the bones" (daah-di-di-daah)
-.--	**YANKEE** "Yankee Doodle" (daah-di-daah-daah)
--..	**ZULU** "Shaka(--) Zulu(..)" (daah-daah-di-dit)

CONTRACTED PUNCTUATION MARKS

.	.-.-.-	The (A)lpha written three times in succession signals that a new sentence is about to start
?	..--..	A request to expand on content or to repeat what was said
,	--..--	The dits between daahs, separate lengthy thoughts
/	-..-.	A slash is a half X, an abbreviated X
"	.-..-.	Enclose xactly what's said, an enclosed (.X.)
;	-.-.-.	The (N)ovember written three times in succession signifies the end of a main thought

CONTRACTED NUMBERS

Dit = one or minus ten Daah = plus five or minus five

CHAPTER 10: HOW TO REVIEW

DEVELOP YOUR OWN CODE SOUND WORD OR PHRASE

In the unlikely event of a particular code sound word or phrase not helping enough for a particular code then search for your own way of reinterpreting the code. For example if you can't remember "Yankee Doodle" for (Y)ankee then feel free to modify a phonetic or code word phrase or to devise your own if necessary. You are by no means limited to just what is listed in this manual. Sometimes nothing can compare to your individual creativity.

If this manual teaches you nothing else, it should at least show that you that the code is not as difficult as it may have first seemed, or as it is commonly believed when you have a method.

REVIEW FREQUENTLY

Frequently means often. Two minutes a day is better than two hours at the end of each month and better still than a whole two days at the end of the year. Schedule a daily routine and adhere to it as much as possible.

It is the frequency rather than the length of practice that counts. The secret is in how often rather than in the total overall time spent. This cannot be stressed enough.

Take frequent breaks during sessions and try not to have a practice session of more than an hour or so. Don't try to learn the entire booklet in one continuous sitting. It will take a few sessions. It might be of some consolation to you that this is already an accelerated course.

THE LEARNING PLATEAU

Keep in mind that when trying to learn anything in general, and in accordance with review and practice that there is a maximum amount that a person can learn in a particular amount of time. When this maximum level or plateau is reached, it becomes difficult to add more to what's already learned. Studying or practicing code is then no longer the best way to spend your time. You'll recognize this when it happens. When it does, it is time to end a session. Don't be discouraged because it's normal.

The second way that we plateau is that it is possible to reach a particular speed both in sending and copying and will seem to get stuck there, finding it hard to break free to a higher speed.

If you are stuck at about 10 wpm, read over "KEEPING WITH THE TRUE SOUND", and "HOW TO HEAR AND RECOGNIZE" in the FAMILIARIZATION section.

TRANSMIT AT 1 MILLION WPM BUT DECODE AT 1 WPM

You will soon discover especially with this course, that you'll be able to send at a faster rate than you'll be able to copy. This too is normal. It is thus a good idea to spend a little more time on copying once you have learned most of the code, but don't completely sacrifice one for the other initially. Consider taking some time to enjoy what you have accomplished by engaging in code conversations within your capability.

Once you become familiar with transmitting the rhythms, you will be able to recognize them without much problem. See "NAME THE TUNE, PLAY THE TUNE" in the THINGS TO KNOW BEFORE YOU START section.

You can learn to decode from sending because when sending you develop a direct association between the character and the code sound that you hear yourself send for the character. Push yourself a little beyond what you're comfortable with but also keep in mind that sometimes you will actually learn more by going slower. Use the code sound phrases and listen for the phrase or word that the code seems to say.

CARRY AROUND THE CODE

Difficult characters should go on to a problem list. Later, make yourself an overview on an index card, and if you find both hands at your side then look them over. Use the methods described in "SILENT CODE MORSE PRACTICE", in the HOW TO LEARN THE CODE SOUNDS AND PRACTICE ANYWHERE section.

CHAPTER 11: Q-SIGNS AND PROSIGNS MADE EASY

Q-SIGNS

Ham radio operators have a standardized system of shortcut abbreviations which are called Q signs because they start with the letter Q. The Q is considered to be a sort of introduction. The three-letter signal can either be a statement, or a question if it followed by a question sign. To make these signals easy to remember we'll just focus and work with the last two letters.

QRG
What's my exact frequency? What's my Rig Generating? What's the exact frequency of? The exact frequency is..

QRL
Is this frequency busy (occupied)? Is this frequency alReady Loaded. This frequency is busy.

QRM
Does my transmission have interference? Your transmission has interference. Can consider interference to be Radio Magnetic interference.

QRN
Is static a problem? Static is a problem. Can consider static to be electrical RaiN.

QRO
Want me to increase power? Should I ROtate my power dial? Please increase your power.

QRP
Want me to RePeal my wattage? Should I decrease my power? Please decrease your power.

QRQ
Should I send faster? Do you Request Quicker? Send faster at x wpm.

QRS
Should a send slower? Do you Request Slower? Send slower at x wpm.

QRT
Should I stop sending? Should I temporarily ReTire?

QRU
Have you anything for me? I have nothing for you. The first person transmits "ZeRU (zero, nothing) for me?" The contact replies, "ZeRU, I've got nothing".

QRV
Are you RiVeted and ready to go? I'm ready.

QRX
RX is the standard electrical abbreviation for receive, receiver or reception. The RX is a way of asking "When will I receive your signal again?" or a way of saying "We'll re-establish contact at..".

QRZ
Who is calling me? I want to RecogniZe the person calling me. I am the person calling you... so you can RecogniZe the person calling you.

QSB
Is my signal fading? Your signal is drifting off into the SideBands.

QSK
Can you hear me and if so can I break in on your conversation? I can hear you and you can break in. This is similar to a CQ (literally seek you) which is transmitted when looking for a station to exchange words with, but this is used when a conversation is already in progress. This is a type of SeeK.

QSL
Can you acknowledge receipt? I acknowledge receipt. Transmitting SL, for SignaL, to acknowledge is as if saying "I got the info(SignaL)". In question form, it's "Do you have the info?" or "Can you acknowledge?"

QSN
Did you hear me on x KHz? Did you hear what I was SayiNg on x KHz? I heard you on x KHz.

QSO
Conversations often start with so. "SO, how's it going?" or "SO, how are the children?". QSO is another word for contact or communication. As a statement it's, "I can communicate with" or "I'm in contact with". As a question it's "Can I communicate with?"

QSP
Will you relay the SPecifics to? I'll relay the SPecifics to..

QST
Indicates that message following is addressed to all Amateurs and ARRL members i.e., "Attention all STations".

QSX
Will you liSten on X KHz? I'm liStening on X KHz.

QSY
Should I transmit on a different frequency? Should I Send to You on x frequency? I am Sending You on x frequency.

QTB
Do you agree with my count of words? Do you request a Trace Back? I do not agree with your count of words. I request a Trace Back. I'll repeat the first letter or digit of each word group. I'll Trace Back.

QTC
How many messages do you have to send? What's The Count? I have x number of messages to send.

QTH
What Town are you in or what Hill are you on? What's your location? My location is..

QTR
What time is it? This is a Time Request.

Q-Sign Overview And Contraction

RG Rig Generating	SB SideBands
RL alReady Loaded	SK SeeK
RM Radio Magnetic	SL SignaL
RN electrical RaiN	SN SayiNg
RO ROtate the power dial	SO Contact, Communication
RP RePeal wattage	SP SPecifics
RQ Request Quicker	ST all STations
RS Request Slower	SX liSten on X frequency
RT ReTire	SY Send to You
RU zeRU	TB Trace Back

RV RiVeted	TC The Count
RX Receive	TH Town or Hill
RZ RecogniZe	TR Time Request

PROSIGNS (PROCEDURAL SIGNS)

Prosigns are one or two letters that represent a phrase or intention. They are commonly studied, and are more popular, as two separate letters even though the code is transmitted as a single unit.

R (.-.)
Message Received. The letter R is transmitted as short for the word "Received".

K (-.-)
Go Ahead. The letter K is short for "OKay go ahead".

KN (-.--.)
Go ahead for specified/Named station. The addition of November indicates that the Go Ahead is meant for the Named station only.

AS (.-…)
Standby (Wait). The three dits after Alpha(.-) seem to say "There's more... stay tuned" or can also consider the AS as short for "Give me A Second".

ov**AR** (.-.-.)
This is the equivalent of the voice communication's familiar "over". It's short for "Over to you" meaning that your contact is now listening and that it's your turn to talk.

SK (...-.-)
End of Contact. CQ's and QSK's are used to open contacts and a contact is closed with an SK which can be considered as short for "Seek Klosed".

CL (-.-..-..)
Short for station CLosed, going off the air.

ERROR (.........)
A long series of dits which all sound the same, is what you transmit when everything starts to sound the same. It means "I'm confused" or "I've made an error". (Not really a prosign, but good to know). Can also be as few as four dits with a repeat of the sequence after a pause.

Prosign Overview And Contraction

R	Received	AR	over (ovAR)
K	Okay, go ahead	SK	Seek Klosed
KN	Go, Named station	CL	Station CLosed
AS	Standby (A second)	Error	....

Made in the USA
Monee, IL
04 November 2026

67374699R00022